where
does
your garden
grow?

BY **AUGUSTA GOLDIN**

# where does your garden grow?

ILLUSTRATED BY **HELEN BORTEN**

 LIBRARY EDITION

# LET'S-READ-AND-FIND-OUT SCIENCE BOOKS

Editors: **DR. ROMA GANS**, Professor Emeritus of Childhood Education, Teachers College, Columbia University

**DR. FRANKLYN M. BRANLEY**, Chairman and Astronomer of The American Museum–Hayden Planetarium

*Air Is All Around You*
*Animals in Winter*
*A Baby Starts to Grow*
*Bats in the Dark*
*Bees and Beelines*
*Before You Were a Baby*
*The Big Dipper*
*Big Tracks, Little Tracks*
*Birds at Night*
*Birds Eat and Eat and Eat*
*Bird Talk*
*The Blue Whale*
*The Bottom of the Sea*
*The Clean Brook*
*Cockroaches: Here, There, and Everywhere*
*Down Come the Leaves*
*A Drop of Blood*
*Ducks Don't Get Wet*
*The Emperor Penguins*
*Find Out by Touching*
*Fireflies in the Night*
*Flash, Crash, Rumble, and Roll*
*Floating and Sinking*
*Follow Your Nose*
*Giraffes at Home*

*Glaciers*
*Gravity Is a Mystery*
*Hear Your Heart*
*High Sounds, Low Sounds*
*Hot as an Ice Cube*
*How a Seed Grows*
*How Many Teeth?*
*How You Talk*
*Hummingbirds in the Garden*
*Icebergs*
*In the Night*
*It's Nesting Time*
*Ladybug, Ladybug, Fly Away Home*
*The Listening Walk*
*Look at Your Eyes**
*A Map Is a Picture*
*The Moon Seems to Change*
*My Five Senses*
*My Hands*
*My Visit to the Dinosaurs*
*North, South, East, and West*
*Oxygen Keeps You Alive*
*Rain and Hail*
*Rockets and Satellites*

*Salt*
*Sandpipers*
*Seeds by Wind and Water*
*Shrimps*
*The Skeleton Inside You*
*Snow Is Falling*
*Spider Silk*
*Starfish*
*Straight Hair, Curly Hair**
*The Sun: Our Nearest Star*
*The Sunlit Sea*
*A Tree Is a Plant*
*Upstairs and Downstairs*
*Watch Honeybees with Me*
*What Happens to a Hamburger*
*What I Like About Toads*
*What Makes a Shadow?*
*What Makes Day and Night*
*What the Moon Is Like**
*Where Does Your Garden Grow?*
*Where the Brook Begins*
*Why Frogs Are Wet*
*The Wonder of Stones*
*Your Skin and Mine**

*AVAILABLE IN SPANISH

Copyright © 1967 by Augusta Goldin. Illustrations copyright © 1967 by Helen Borten. All rights reserved. Except for use in a review, the reproduction or utilization of this work in any form or by any electronic, mechanical, or other means, now known or hereafter invented, including photocopying and recording, and in any information storage and retrieval system is forbidden without the written permission of the publisher. Published in Canada by Fitzhenry & Whiteside Limited, Toronto. Manufactured in the United States of America. Library of Congress Catalog Card No. 67-18517.

REC Library Edition reprinted with the permission of Thomas Y. Crowell Company
Responsive Environments Corp., Englewood Cliffs, N. J. 07632

ISBN 0-690-88357-9
0-690-88358-7 (LB)

# Where does your garden grow?

LET'S READ AND FIND OUT →

There is soil under the grass and under the trees. There is soil under roads and sidewalks, rivers and lakes. There is soil at the bottom of the sea.

Unless your house is built on solid rock, there is soil under your house.

Soil is under your feet when you walk in the country.
It is under your feet when you play in a city lot.

Strawberries and watermelons grow in topsoil. Apples and pears, cherries and bananas grow in topsoil.

You may see loose, damp soil around bushes and trees. You may see a beautiful garden, planted in rich, loose soil. Rich, loose soil is called TOPSOIL.

Take a walk and look at the soil where you live.
You may see rough, rocky soil full of gravel and pebbles.
You may see sandy soil.
You may see smooth, hard soil.

Almost
all the food you eat
grows in topsoil.
Even meat and milk,
cheese and eggs
come from topsoil.
They come from
animals
that eat the plants
that grow in topsoil.

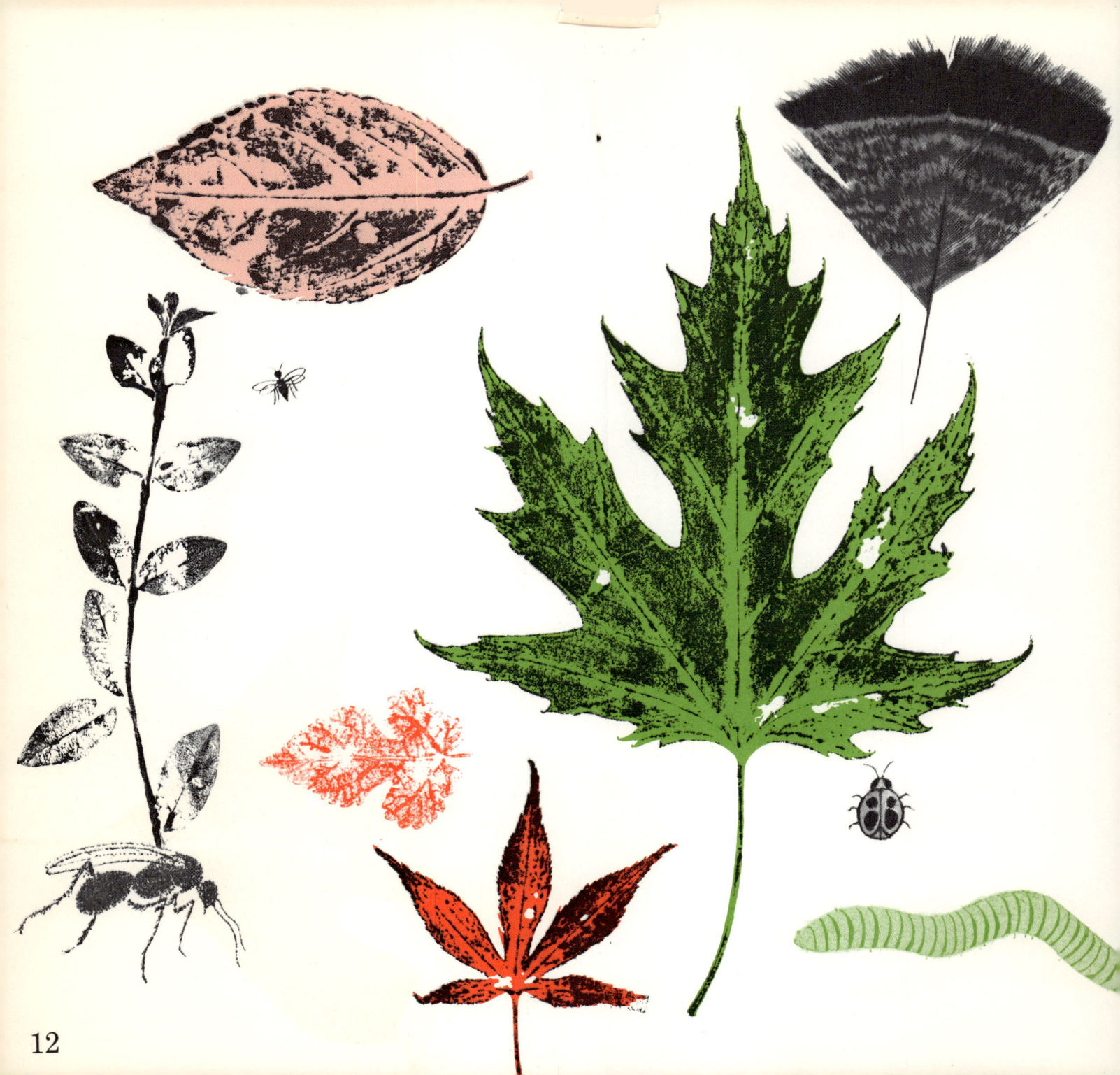

Fruits and vegetables, trees, flowers and grass grow well in topsoil because topsoil contains HUMUS. Humus keeps the soil loose, and helps it hold water. It makes the soil a good place for plants to grow.

Humus is made up of dead leaves and twigs, of dead bugs and worms, of small animals and plants. When humus is mixed with soil, it makes topsoil.

The humus soil you buy in a store has been ground up. It looks like smooth, dark powder.

The humus in fields and gardens is rough.

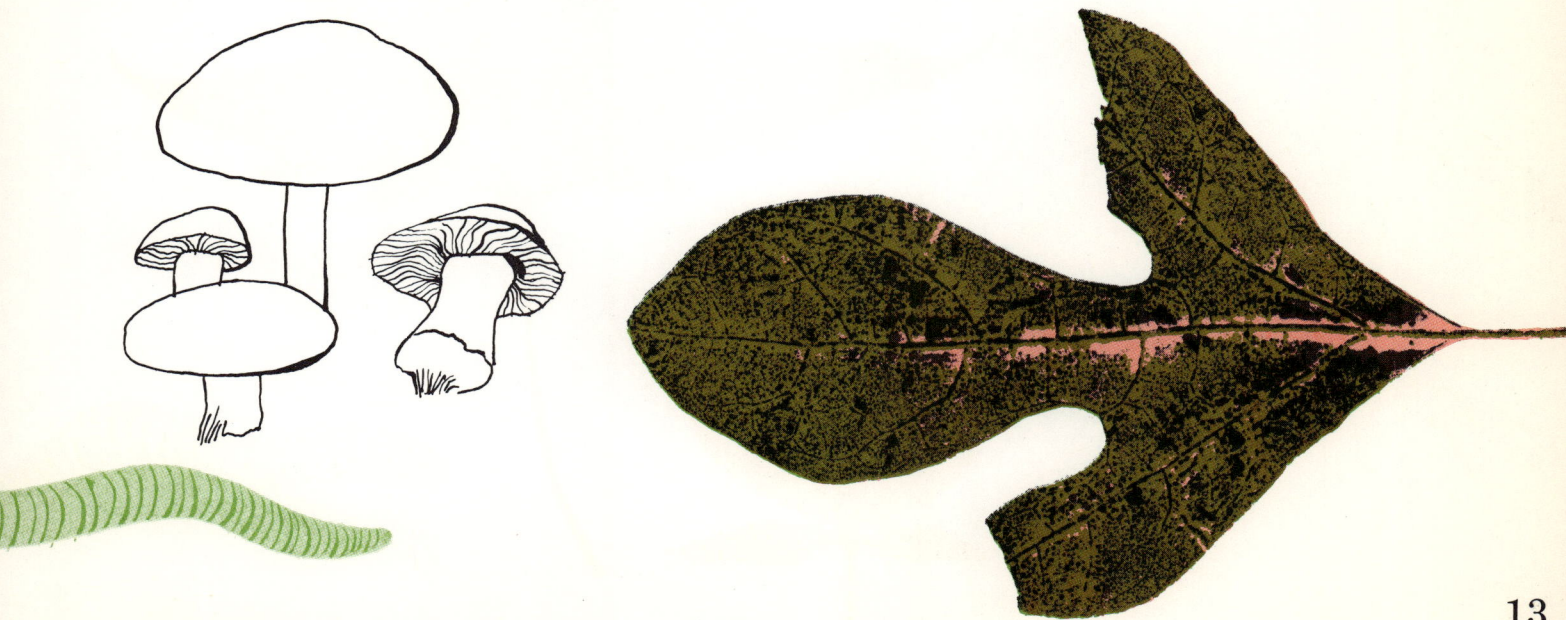

Rake some leaves that were piled up in a field or lot during the winter and you will see the beginning of humus. The leaves underneath will be wet and slippery and black. They will be mixed with twigs, dead bugs, and plants too small to see.

After a while, these leaves and twigs, these animals and plants will decay. Then they will be humus.

In the fields and woods, humus is mixed into the topsoil, every year. Some of this humus comes from the weeds and grasses and leaves that grow *on* it. Even more humus is mixed into the topsoil by the millions of plants and animals that live *in* it. Topsoil is a home for living things.

You can find this out for yourself.

If you live in the country, go to a woods where the soil is damp and spongy. If you can't find a woods, go to a shady brook or pond. Look carefully at the soil. Poke it with a stick. You will see ants and worms and beetles squirm and run away. If you dig in the soil, you may find bits of dead beetles, snails, insects, and the bones of small animals. You will also find bits of roots and stems mixed in with the soil.

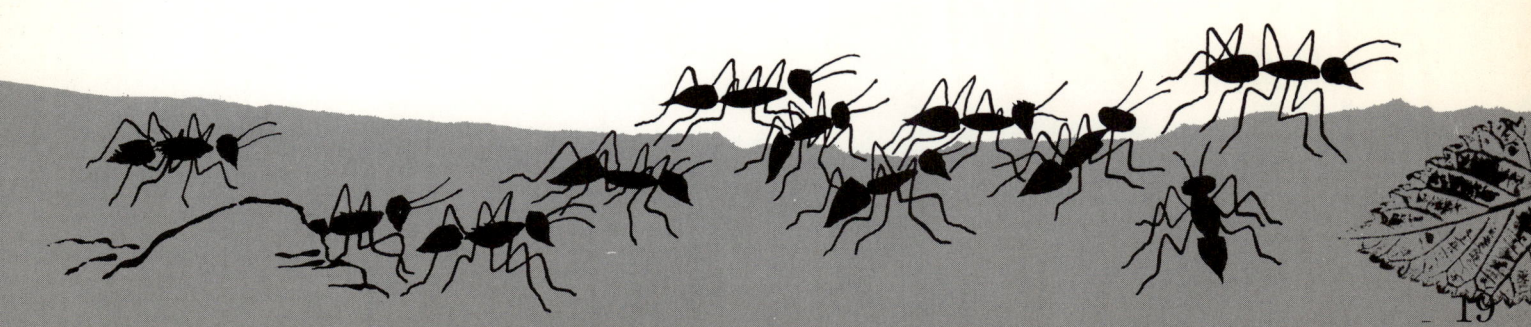

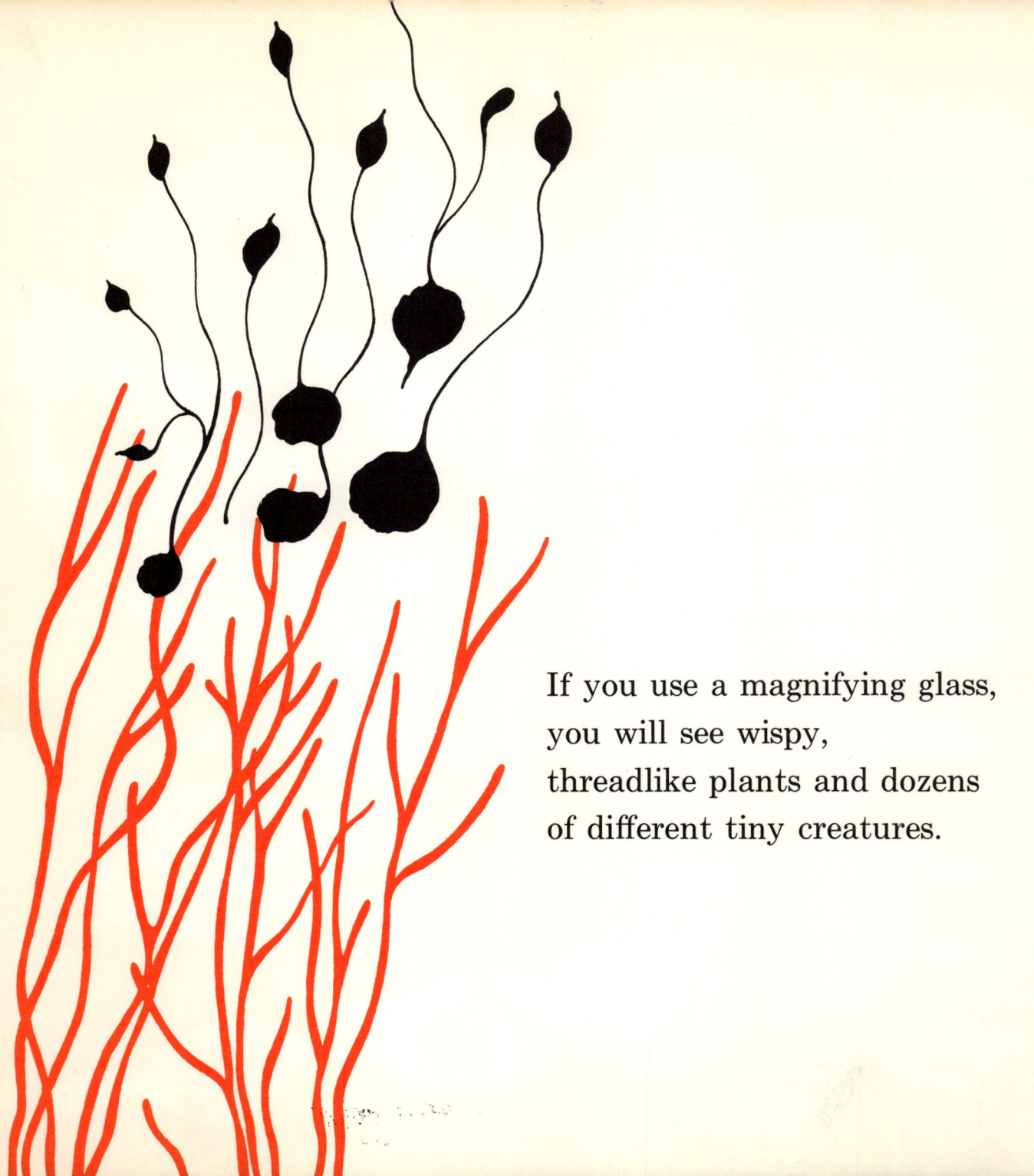

If you use a magnifying glass,
you will see wispy,
threadlike plants and dozens
of different tiny creatures.

There are thousands of other plants and animals in the soil, so small you would need a microscope to see them.

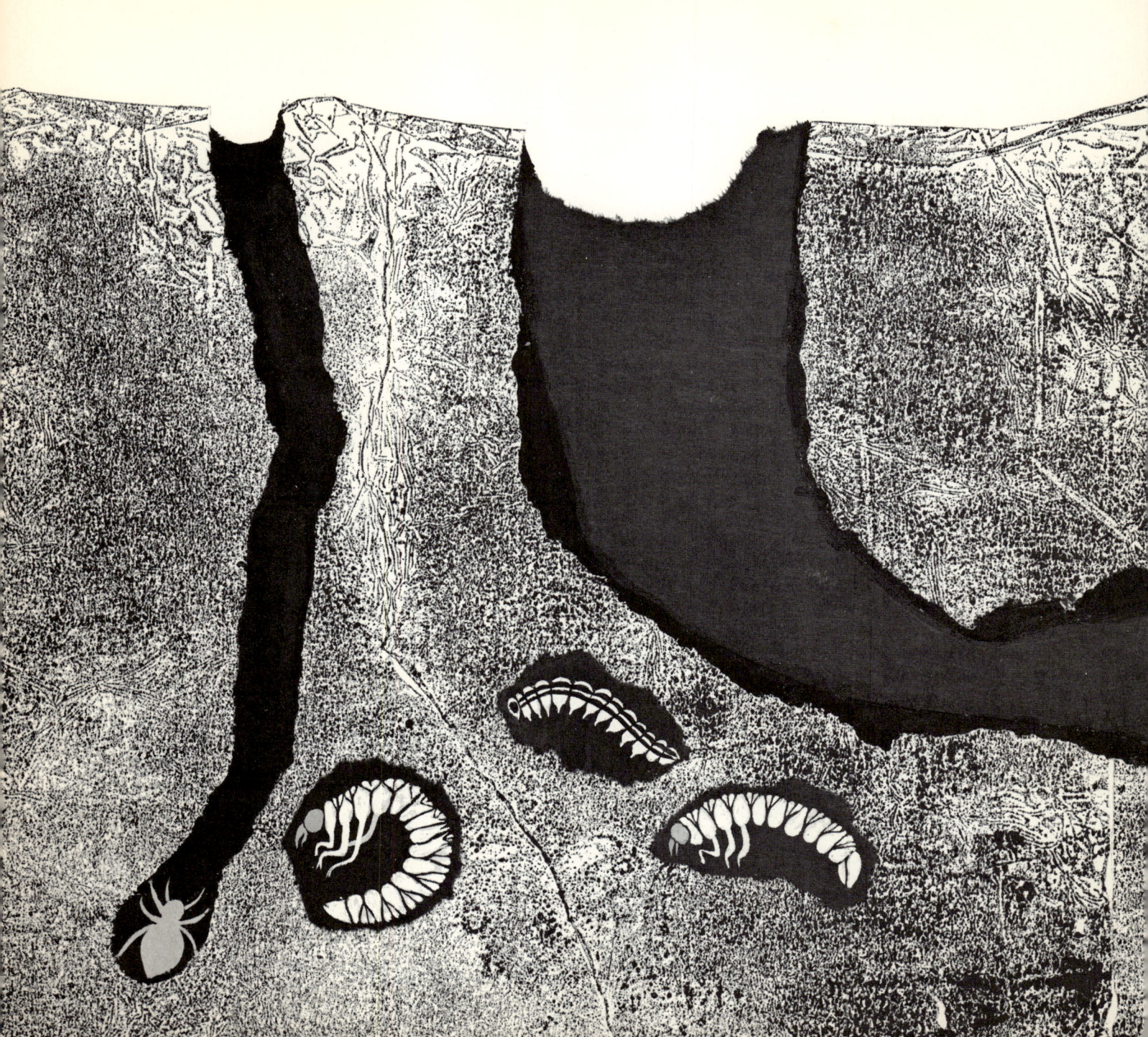

Topsoil is crowded with plants and animals that live underground.

When they are alive, underground plants and animals feed on the humus in the topsoil.
When they die, underground plants and animals decay and also turn into humus. The humus mixes with the soil to make more topsoil.

If you live in a city, go to an empty lot and look at the soil. It will be hard and dry. There will be very little humus in it. You'd be disappointed if you tried to garden in city soil. Strawberries and watermelons would not grow well without humus. Corn and beans would grow thin and straggly. Fruit trees would never get big enough to bear fruit in hard-packed city soil that has no humus in it.

It is good that there is topsoil. It is good that plants and animals live in the soil and on the soil. It is good that plants and animals die and decay and turn into more humus. It is good that humus mixes with soil to make topsoil.

There is not as much topsoil in the world as you might think. In a few places, the topsoil is a hundred feet deep, but there are not many places like that. On many plains, the topsoil may be only two or three feet deep. On hills and mountains, it may be only two or three inches deep.

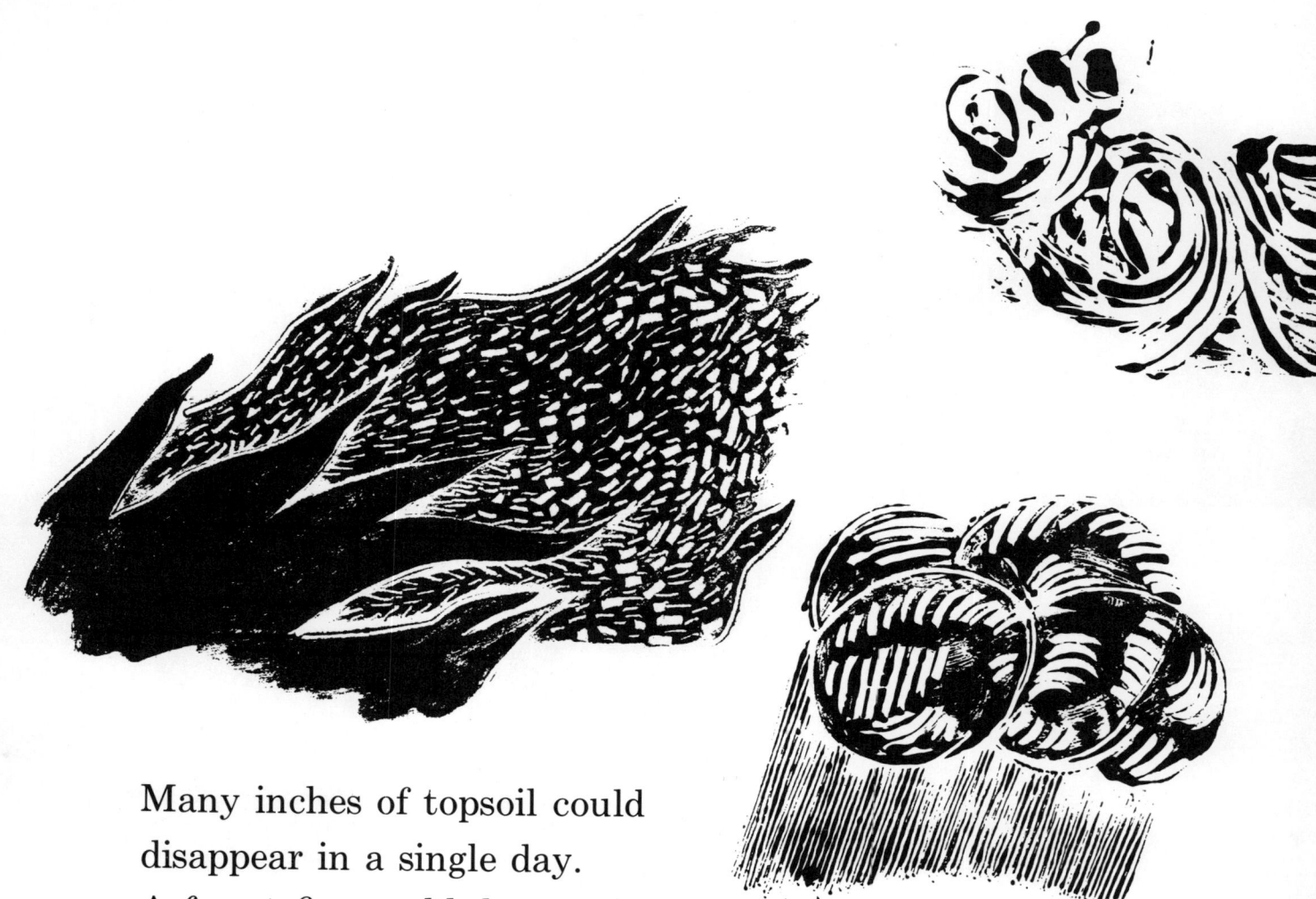

Many inches of topsoil could
disappear in a single day.
A forest fire could destroy it.
A hard rainstorm, or a flood,
could wash it away.
A dust storm could blow it away.
A tornado could pick it up and whirl it away.

As long as topsoil stays where it belongs, that's where your garden will grow. Then there will be good things to eat—carrots and peas—corn, beans, and pumpkins—strawberries, watermelons, apples and pears, and cherries.

## ABOUT THE AUTHOR

Augusta Goldin was born in New York City but grew up on a farm in the Catskill Mountains near Ellenville, New York. She was graduated from Hunter College, received a Master of Science degree from the City University of New York, and a Ph.D from Teachers College, Columbia University.

Mrs. Goldin has worked on the staffs of several educational publications, and is the principal of a school on Staten Island, New York. She has written several previous Let's-Read-and-Find-Out Science Books.

## ABOUT THE ARTIST

Helen Borten has illustrated many books for children and has written and illustrated several others, including HALLOWEEN, a Crowell Holiday Book.

Her illustrations have been cited by the American Institute of Graphic Arts and by *The New York Times* in their annual list of ten best-illustrated books.

Mrs. Borten was born in Philadelphia, Pennsylvania, and was graduated from the Philadelphia Museum College of Art. She lives with her family in Lafayette Hill, Pennsylvania.